Reading Essentials
in Science

HOW THINGS ARE MADE

TEXTILES
Smooth as Silk, Bumpy as Burlap

BETH DVERGSTEN STEVENS

PERFECTION LEARNING®

Editorial Director: Susan C. Thies
Editor: Judith A. Bates
Design Director: Randy Messer
Book Design: Tobi S. Cunningham, Lori Gould
Cover Design: Mike Aspengren

Dedication

This book is for kids who like to make things with their hands. May you discover new talents as you play with fibers, yarns, and fabrics.

Acknowledgements

I wish to thank the following experts for sharing their time and knowledge about fibers and fabrics so this book could be written.

Grace and Oliver Johnson, Waverly, Iowa
Marty Olson, Carla Bowlin, Alice Bullers, and Carol Dahms, Northeast Iowa
    Weavers and Spinners Guild
Sonny Driggars, Hanes Manufacturing, Sarah Lee Sportswear
Jan Stone, Iowa State University, Textiles and Clothing Department

A special thanks to Kristin Mandsager, instructor of physics and astronomy at North Iowa Area Community College, for his scientific review of the book

IMAGE CREDITS
©Ted Horowitz/CORBIS: p. 10 (top); ©Yann Arthus-Bertrand/CORBIS: p. 28; Mary Evans Picture Library: p. 30; Photodisc Green: pp. 17, 19 (bottom), 20, 26; Rubberball Productions: p. 25; Beth Dvergsten Stevens: pp. 8, 12, 13 (bottom), 14

Photo.com: cover (bottom left and bottom right), back cover, pp. 4, 5, 7 (top), 9, 10 (bottom), 11, 16, 23, 24 (top), 24 (bottom), 27, 31, 32, 36, 37, 40; Corel: cover (top and bottom middle), p. 29; Digital Vision: p. 1; Liquid Library: p. 6; Perfection Learning Corporation: pp. 3, 7 (bottom), 13 (top), 15, 18, 19 (top), 22, 33, 34, 35

Phone: 1-800-831-4190
Fax: 1-800-543-2745
perfectionlearning.com

1   2   3   4   5   BA   07   06   05   04   03

Paperback ISBN 0-7891-6123-0

Table of Contents

History and Development of TEXTILES

You wear it, sleep on it, and dry off with it. It can keep you warm or cool. It can be strong enough to stop bullets or so thin you can see through it. What is this amazing material? **Textile**!

There are many different textiles today. They are made in special ways to do different jobs. The denim in your blue jeans needs to be tough but comfortable to wear. The terry cloth in your bath towel needs to be soft and absorbent. Kevlar in a bulletproof vest is very different from satin in a wedding gown.

Yet all textiles have something in common. They are "built" from tiny **fibers**. The type of fibers and the way they are put together makes each textile different.

Some fibers come from nature. Plants, grazing animals, and caterpillars give us **natural** fibers. Five common ones are cotton, linen, jute, wool, and silk.

Other fibers come from chemicals. These are **synthetic**, or humanmade, fibers. Most of the chemicals begin as **crude** oil from deep underground. Scientists have learned how to turn chemicals into fibers, such as polyester, acrylic, and nylon.

Natural Fibers from Plants

Cotton

Cotton comes from the seedpods of cotton plants. These plants grow best in warm, sunny places such as Texas. The plants grow for five or six months before any cotton can be harvested.

Two months after cottonseeds are planted, flowers bloom on the plants. After they die, seedpods, or cotton **bolls**, are left behind. White cotton fibers grow inside

each boll. The bolls grow bigger and bigger until they turn brown and pop open.

Sunshine and warm air help the cotton fibers dry out and become fluffy. When nature's process is finished, the bolls are ready to be

Cotton fields

picked. Long ago, cotton was picked by hand. Today, big machines pick the bolls.

Many seeds are tangled in the cotton fibers. A cotton gin removes the seeds. These seeds are sold to a factory to make cooking oil.

Machines clean the cotton fibers. Then the cotton is packed into huge bales and inspected. The bales are sold to cotton merchants and mills. In the mills, machines spin the fibers into yarn. Other machines weave or knit the yarn into fabrics. The fabrics are then sold to sewing factories that make clothing and other cotton items.

You probably use cotton every day. Cotton is the most common fiber. Blue jeans, towels, T-shirts, sweatshirts, bedsheets, socks, and money are all made from it. Why? Because cotton absorbs moisture well. The fibers are soft and light. It's cool and comfortable to wear.

It's not very expensive either.

Linen

Linen comes from the **flax** plant. The fibers hide inside the thin, 3-foot-tall stalks. To remove the flax fibers, the plants are pulled out of the ground by their roots and placed in water to soak, or **ret**. This process softens the stalks.

Once the stalks have dried, the bark on the outside of the stalks is scraped off. The **pith** inside is scraped off the fibers by blades on a scrutching machine. Then the fibers are combed and spun into yarns and woven into linen fabric.

Since flax fibers are smooth,

linen doesn't shed little fiber pieces, or **lint**. So linen is good for dish towels, tablecloths, and clothing. But linen fabrics wrinkle easily.

Jute

Jute plants are very tall—as tall as two men! The fibers are harvested from the stems, much like flax. It's a rough, brown fiber that isn't very pretty. But it's cheap and useful. Jute is used for burlap bags, carpet backing, and twine.

Burlap

Natural Fibers from Living Creatures

Wool

Wool is made from fleece that grows on sheep. Each spring before the weather gets too warm, workers use clippers to **shear** a layer of wool from each sheep. It's like a haircut. The wool is cut off in one piece.

Sheep

The fleece grows back in time to keep the sheep warm during the winter. Then the next spring, the sheep are sheared again.

Wool has dirt, weeds, and other barnyard stuff tangled in it. People don't want that in fabric, so the sheared wool is washed.

Lanolin

Lanolin is grease in sheep's wool. It's used to make lotions and makeup. If wool is washed in cold water, lanolin remains in the wool. You can feel it with your fingers. Most wool is washed in hot water to remove the lanolin, which is sold. Lanolin-free wool spins easier and takes dye better.

Carding wool

In factories, machines spin rolags into yarn. Then other machines use the yarn to make woolen textiles.

Some people still spin their own yarn by hand. First they pull fibers from one end of the rolag. The fibers slide by one another. This is known as **drafting**. The fibers are spun on a spinning wheel where they are twisted together into yarn. The yarn is wound on big **bobbins**.

Wool is used to make sweaters, blankets, coats, and suits. Wool clothing keeps people warm and dry because the fibers are like tubes with scales. They hold air and keep moisture away from the skin.

Wool must be combed, or **carded**. This takes out the tangles, lines up the wool fibers, and makes the wool soft and smooth. Carded wool fibers are shaped into a loose roll, or **rolag**.

Spinning wheel

Other animals, such as llamas, alpacas, rabbits, and goats, have hair used to make woollike fabrics. Their undercoats make very soft yarn.

Hair from alpacas can be 12 inches long! It's very soft and straight. Fibers are thick and hollow. They can be spun into yarn much like sheep's wool. Clothing made from these yarns is very warm but not too heavy to wear.

Silkworms

Silk

You've probably seen caterpillars and moths in your yard. There are thousands of different kinds. But one caterpillar is very special—the silkworm. It spins a cocoon from a long, fine thread called *silk*. When the cocoon is unwrapped, the shiny thread is about a mile long!

Farms in China and Japan raise silkworm moths. One moth can lay 500 eggs. A tiny silkworm hatches from each egg.

Then the silkworms go to work. The farmer sets them on piles of mulberry leaves. The silkworms eat leaves almost nonstop. Before long, their skins get too small. So they molt, or shed, their old skin and eat it. A caterpillar molts four times as it grows.

When the caterpillars stop eating, the silk farmer puts them in a cardboard frame. There is a small space for each silkworm. Each one sends out a thin strand of liquid from its body. This makes a messy web that holds the silkworm in the frame.

Each silkworm then begins to spin a neat cocoon around itself. As the liquid silk comes out, it hardens in the air and becomes a long silk fiber. The silkworm needs about three days to spin a whole cocoon.

The silkworm grows into an adult moth inside its cocoon. When it's finished growing and changing, it makes a hole in the cocoon to get out. That breaks the silk thread. But silk farmers don't want that to happen. They want long, unbroken threads.

So the farmers only let *some* of the caterpillars grow into moths. The rest of the cocoons go into an oven where the caterpillars die inside the cocoons. This is called *stifling*.

Silkworm cocoons

After stifling, the cocoons are soaked in hot water. That loosens the silk fibers so the cocoons can be unwrapped. Five to ten silk fibers are pressed together into one strong silk thread. One thread is the size of a hair on your head, but it's strong enough to hold a pound of butter!

Silk is shiny and smooth. People buy blouses, scarves, stockings, and ties made from silk. Silk is strong enough for parachutes and ropes too.

Synthetic Fibers

Synthetic fibers come from the same chemicals that give us gasoline. Scientists just put them together differently to turn them into fibers.

Spools of synthetic fibers

How? Chemicals are cooked together in big tanks. The mixture is poured onto a roller, dried, and cut into flakes. The flakes are melted and pushed through a nozzle with lots of little holes. This is a **spinneret**. The long fibers that come out of each hole are called *filaments*. They're almost like cooked spaghetti, except much longer.

When air hits the filaments, they harden. The long fibers are wound onto a spool. Then they are stretched to make them stronger. Finally they are rolled on a larger spool, ready to be used.

Some filaments are cut into short pieces. This makes them look more like cotton or wool fibers. Others are long and uncut. These fibers look more like silk. Synthetic fibers are spun into yarns and made into fabrics at a **mill**.

Polyester

Polyester is the most common synthetic. It is often mixed with other fibers to improve them. For example, polyester helps cotton last longer and wrinkle less.

Acrylic

Acrylic is a wool "copycat." Peek at the label inside a sweater. Even if the sweater looks like wool, it might be made of acrylic. Acrylic and wool fibers are sometimes blended together.

Nylon

Nylon was the first synthetic fiber to be made from just chemicals. This strong, durable fiber is used for lightweight things such as stockings and swimsuits. It's also used for big things like sleeping bags, parachutes, carpeting, and furniture.

Every year, scientists figure out how to make new and better synthetics.

Nylon tent

11

2

From Fiber to Yarn to TEXTILE

From Fiber to Yarn

Before textiles can be made, fibers must be turned into yarn. When fibers are gathered, they are tangled. To get the tangles out, the fibers are pulled across small wires, like bristles in a brush. This is called *carding*. Soon they run the same direction.

A mill might mix different fibers together during this step. For example, cotton could be mixed with polyester.

Since fibers are so fine, many

Dropping the spindle so the twist will travel and make more yarn

must be twisted together by spinning machines. This makes a long, strong yarn. Yarns can be soaked in big vats of dye to make them colorful.

Long ago, people carded wool with simple hand tools. They twisted the fibers into yarns with their hands or with a drop **spindle**. The invention of the spinning wheel helped people make yarns faster. Some people still enjoy using hand tools and spinning wheels to make yarn today.

Spin Yarn the Old-Fashioned Way

Alice Bullers teaches Iowa children to spin yarn by hand using clothespins. Try her technique.

1. Using wool or llama hair, twist some fibers between your fingers. Clip them inside the clothespin.
2. Draft, or pull, the fibers for three inches with your left hand, and then pinch them.
3. Lightly roll the clothespin down your leg many times with your right hand to twist the fibers together.
4. When they are very tight, put the clothespin between your knees. Move your right hand to pinch the fibers where your left hand was.
5. Draft another 2 inches of fibers with your left hand. Then let go with your right hand. *Boing!* Watch the twist move up the drafted fibers toward your left hand.
6. Move your hands two more times until you have 6 inches of yarn. This is how drop spindles and spinning wheels work. Wrap the new yarn around the clothespin, and then twist some more.

Clothespin spinning

The bobbin of yarn fits in the center of the shuttle.

The weaver passes the shuttle between warp yarns.

The weaver pulls the beater forward to push the weft yarns together to make fabric.

From Yarn to Textile

Yarns are turned into fabric in three basic ways—weaving, knitting, or felting.

Weaving

People learned how to weave yarns into fabric before they learned to knit. Weaving crosses two sets of yarns. Long ago, people used a wooden hand **loom**. Now, most weaving is done on big factory looms.

On any loom, one set of yarns is called the **warp**. Warp yarns are stretched and attached to the loom. These yarns are as long as the length of the fabric being made. If the warp yarns are 10 yards long, the finished piece of fabric will be just a little shorter than that. The warp yarns are wrapped on a warping board to make them the right length.

The other set of yarns crosses over and under the warp yarns. These are called *filler*, or **weft**, yarns. They are carried through the warp yarns with a **shuttle**. A **beater** pushes the weft yarns together to make the fabric.

Weavers cross the yarns to make many patterns. The

plain weave   twill weave   silk

simplest pattern is the plain weave. This makes a strong fabric. The weft yarns go over one warp yarn and under the next one. This over-under-over pattern is done in reverse on the next row—under-over-under. Look closely at burlap. You'll see a plain weave.

Twill weaves are also strong. The weft yarns go over two warps in a staggered pattern. If you look at twill fabric, you'll see diagonal lines in the weave. Denim is a twill weave.

Some fabrics use a satin weave. The weft yarns go over many warps. If shiny yarns are used, the fabric is smooth and slippery.

Pile weaves have loops of extra yarns. Look at a terry cloth towel and find the loops. Sometimes the loops are cut to make a **plush** side. Corduroy and velvet have plush surfaces.

Some woven fabrics are brushed on one side, like the cotton flannel in pajamas. The brushed side feels soft because the fibers are fluffed up.

There is no limit to the fabrics and patterns that can be made!

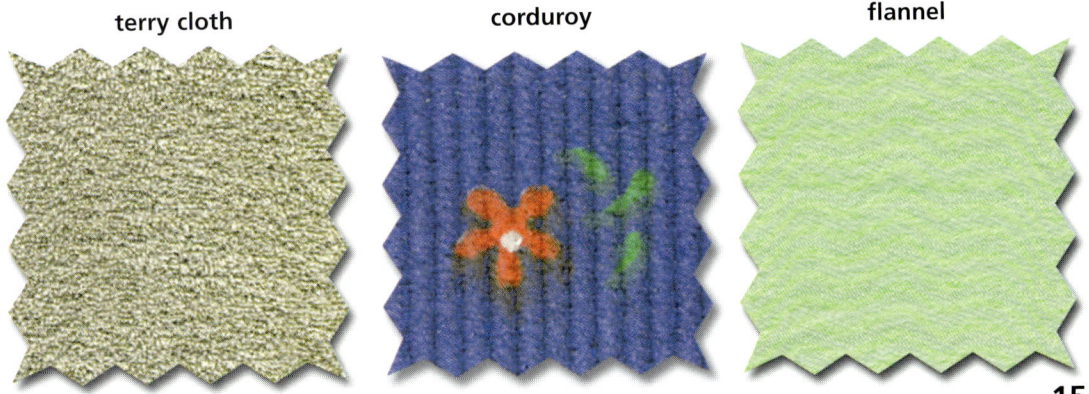

terry cloth   corduroy   flannel

Knitting

Knitting uses knitting needles to make many loops with one single yarn. The loops go through themselves in row after row. When knitting needles are big, the loops are large and loose. Tiny knitting needles make small, tight loops.

Knitted cap and mittens

If you pull on the right yarn in a knit fabric, the whole piece might unravel! Have you seen "runs" in women's pantyhose? That's where the loops pulled away from one another.

Knit fabrics are stretchier than woven ones because loops have "give." Sweaters, T-shirts, sweatshirts, and socks are usually knitted. That makes them comfortable.

Many people still knit sweaters and scarves by hand. But most of your clothing is made on large knitting machines in factories.

Felting

Some fibers are turned into textiles without weaving or knitting. The fibers are tangled up and pressed together with chemicals and heat. Felt, some blankets, and some indoor-outdoor carpeting are nonwoven fabrics.

Finishing the Textile

After a textile is made, it is washed. Cotton is often washed in hot water to preshrink it. When clothes are sewn from preshrunk fabric, they won't get smaller when you wash and dry them at home.

Colors are added to textiles too. If the yarns weren't already dyed, the whole piece of fabric can be dyed or bleached. It will look the same color on both sides. Another way to add color is by printing. When a pattern is printed on one side of the fabric, not all the dye goes through it. The back side looks lighter.

Textile finishes are important. They can be wrinkle-resistant, waterproof, or flame-retardant. These types are specially treated with chemicals.

Everyday
CLOTHES

3

Objects of Interest: Jeans, T-Shirts, Sweatshirts

Look around. What are people wearing? Jeans, T-shirts, and sweatshirts are popular clothes today. All three are made of cotton. But each one feels and looks different. That's because T-shirts and sweatshirts are made by knitting. They are stretchy. Jeans are woven and stretch less.

Jeans

A storekeeper, Levi Strauss, sold twill fabric that was very strong. In 1872, a tailor named Jacob Davis bought this fabric to make work pants for a woodcutter. Cutting wood was hard work.

The man's pants had to be very strong. Soon other people saw these "waist overalls" and wanted them too.

17

Davis and Strauss teamed up to make twill pants for woodcutters, miners, and cowboys. But sometimes the pockets ripped when miners put ore samples in them. Davis fixed that problem. He put copper rivets at the pocket corners. The pants were called Levis, or blue jeans.

As years passed, jeans became more and more popular. Just about everyone owns a pair today.

Jeans are made from denim fabric. Denim is woven in a mill from white and blue yarns. The textile is wrapped on large bolts that weigh about 500 pounds. The denim is sent to blue jeans factories. The bolts are unrolled, and a hundred or more layers of cloth are stacked up.

Workers follow a pattern to cut the denim into pieces. Electric saws help them cut through a whole stack of fabric at once.

On huge sewing machines, other workers sew the denim pieces together in the right order. A single pair of jeans needs more than 30 sewing steps before it is done. More than 200 yards of thread are used.

Look at a pair of jeans. You will see seams, pockets, zippers, belt loops, and hems that were sewn.

Rivets

Metal rivets made pockets stronger. But when people sat down, the rivets scratched furniture. Cowboys had another problem. When they stood by a fire, the rivets got very hot. Ouch! Since the 1960s, many jeans have been made without rivets. Extra stitching has been added at stress points.

Explore Denim

Unravel the cut edges of a denim scrap. Look for the white and blue yarns. The blue yarns run lengthwise. They were the warp yarns on the loom. The white yarns were the weft yarns, woven through with a shuttle.

Now use a magnifying glass to look at the twill weave pattern. The yarns are woven tightly to make denim strong.

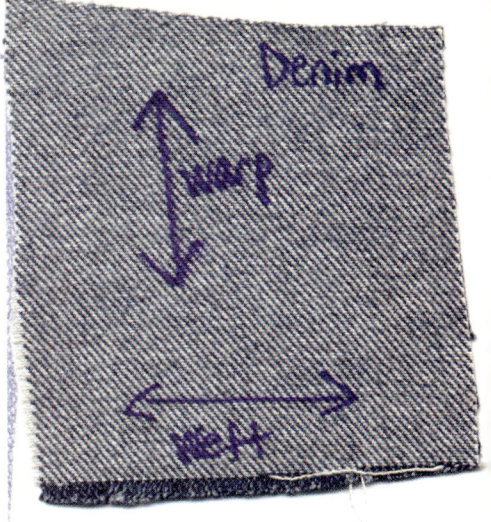

T-Shirts

T-shirts and sweatshirts are both knitted from cotton yarns. But sweatshirts are thick and fuzzy inside. That's because the fabrics are made differently before they are sewn into shirts.

T-shirts are made from plain jersey knit. After a company buys and spins cotton into yarn, the workers knit it into fabric with circular knitting machines. The machines knit a long fabric tube without any seams.

Each machine is a different size to make different-sized shirts. One machine knits jersey tubes for small-sized shirts. Another machine knits larger tubes of fabric for larger shirts.

T-shirts have a ribbed collar around the neck. The ribbing is made on a different machine. Ribbing is stretchier than jersey. That helps you get the shirt over your head!

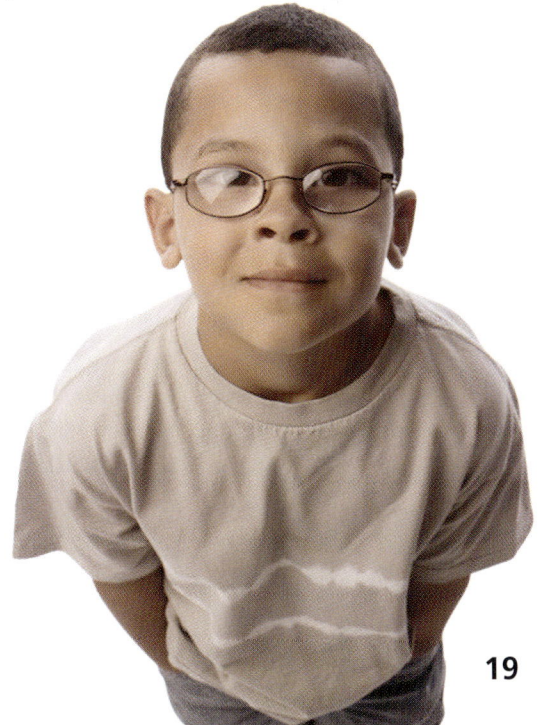

The jersey fabric and ribbing are dyed at the same time. Then they are the exact same color. There are 2000 pounds of fabric in one batch, or **dye lot**.

After other finishes are added, the fabric is preshrunk and dried. Then it goes through heated rollers to make it just the right size.

Four layers of tubular jersey are stacked on a conveyor. A cutter with sharp blades comes down and slices through all the layers at once. It's like using a cookie cutter on dough.

Each T-shirt needs a body piece, two sleeves, and the ribbed collar. The sewers put the shirts together, starting at the shoulder seams. The collar with the label goes on next. The hemmed sleeves are sewn on last.

Look at a T-shirt. Do you see all the seams and sewing steps? Inspectors look closely to be sure the shirts are made well.

Sweatshirts

Sweatshirts are sewn in a similar way. But the fabric is fleecy inside. Sweatshirt fabric is made with lots of loops on one side. A napping machine with sharp wires breaks the loops. The broken loops feel soft and fuzzy.

Copying
NATURE

Objects of Interest: Velcro, Nylon Stockings, Fleece

Nature's plants and animals give people ideas for new inventions. That's how Velcro, nylon stockings, and fleece fabrics were invented.

Velcro

Have you ever walked through a field full of weeds? If there are cockleburs in that field, you'll bring some home with you. Cockleburs are weeds that stick to clothes.

In 1948, a Swiss man named George de Mestral found some cockleburs on his clothes and his dog. They were hard to pick off. He decided to find out why they were so pesky. He put one under a microscope. He saw many little burrs, or hooks, sticking out from the center. Those hooks helped the cocklebur grab and stick to fur and loops in fabric.

De Mestral thought that there must be a way to use this idea as a fastener. He experimented for years. Then he finally did it! He found a way to make matching strips of fabric that hold tight like cockleburs but that come apart easily. One fabric strip had small fuzzy loops like clothes, and the other side had many small stiff hooks like the cocklebur. When the two strips were pressed together, they held tightly. But they could be pulled apart and reused.

Use a magnifying glass to look at a cocklebur. Do you see all the little hooks? Press a piece of fabric against it. Now have a friend hold the magnifying glass. Pull the cocklebur slowly away from the fabric. Can you see how it grips the fabric?

De Mestral named his hook-and-loop fastener Velcro. Today, Velcro is often used instead of buttons, zippers, and laces. You can find it on shoes, in NASA space shuttles, and even in artificial hearts.

What Does Velcro Mean?

The word *Velcro* is a combination of two words—*velvet*, a soft, plush fabric, and *crochet*, a needlework that uses a special hook to make a series of loops.

Nylon Stockings

Before 1900, women's stockings were made from cotton or wool. They were baggy and thick. Women wore long skirts, so it didn't matter what their stockings looked like.

Then in the 1920s, skirts became shorter. Women wanted nicer stockings. Silk stockings seemed like a good answer. They were lightweight and sheer, but they snagged and ran easily. Also, silk was expensive. Many women couldn't afford to buy silk stockings.

Scientists wanted to make artificial silk. First they watched silkworms eat mulberry leaves. Then they observed the long strands of silk that shot out of the silkworms' bodies when they made cocoons.

These scientists decided to make their new fiber from trees, just like the silkworm did. They tried pine, spruce, hemlock, and mulberry trees. After many experiments, they learned how to do it.

They chopped wood into little pieces. Then they cooked the pieces with chemicals and water until the wood pieces were mushy. They bleached the mushy pulp and dried it in sheets. The sheets were soaked in more chemicals, shredded, and turned into a liquid

Nylon stockings

that looked like honey. It was called *viscose*.

The scientists pushed the viscose out tiny holes in a spinneret. The long strands of fiber hardened in acid. It worked! The fibers could be spun into yarns. The yarns could be knit into stockings and other clothing. This first synthetic fiber was named *rayon*.

After scientists learned to make rayon, it didn't take long before they made the first true synthetic fabric—nylon. They made it completely from chemicals, not trees.

Nylon was better than silk or rayon because it was strong, dried fast, and kept its shape better. Pantyhose are still made from nylon today.

Fleece

The wool fleece from sheep is fluffy, warm, and soft. After a sheep is trimmed, the fleece grows back.

Polyester fleece was made to look like wool fleece, but large amounts of crude oil were used in the process. In 1993, textile makers discovered a way to save oil and make fleece fabric. They used recycled plastic soda bottles!

Companies collect thousands of plastic bottles and chop them into tiny flakes. The flakes are cleaned, melted, and pushed through a spinneret to make long fibers.

Bales of fibers are sent to mills. There, the fibers are knit and dyed to make a soft, fluffy fabric. Synchilla and Polarfleece are two common names for this "recycled" fleece.

It takes 3700 2-liter bottles to make enough Synchilla for 150 items of clothing. That's lots of plastic. But that saves 42 gallons of crude oil and creates less air pollution.

Mountain-climbing clothes, hats, and jackets are made from these fleecy fabrics. They keep people warmer than real sheepskin and cost less.

Tough STUFF

Objects of Interest: Flags, Carpeting, Rope, Backpacks, Fireproof Suits, Bulletproof Vests, Swimwear, Sun-Resistant Clothing

Flags, Carpeting, Rope, and Backpacks

Some fibers must be extra strong and tough. Flags that whip around in the wind must not rip or rot in the sun. Carpeting shouldn't wear out when people walk on it. Ropes that tie ships to docks can't break easily. Even the backpacks you carry must be strong enough to hold many pounds.

Flags, carpeting, rope, and backpacks are often made from synthetics such as nylon. That's because those synthetic fibers are very strong. Items made from nylon show less wear than those made from natural textiles. Nylon can be light so flags can move in the breeze. But it can be heavy enough to be used for **artificial** turf on football fields.

Fireproof Suits and Bulletproof Vests

Sometimes clothing has to protect people. Firefighters and race-car drivers need coats and pants that won't burn. Police officers need bulletproof vests. But cotton and wool are soft. They can catch fire. Nylon and polyester will melt. Plus, these fibers won't keep people safe from bullets.

In 1965, a scientist discovered the answer while trying to invent a fiber to replace steel. It had to be stiffer than nylon and not stretch. The result was called *aramid*.

Kevlar and Nomex are two types of aramid fibers. They are very strong and fire-resistant. In fact, one pound of Kevlar is five times stronger than one pound of steel!

Kevlar and Nomex are used in suits for firefighters and race-car drivers. Nomex is worn by factory workers and people who work on electrical lines. The suits protect them from fire and heat.

Cables on navy ships, bulletproof vests, and some work gloves are also made from Kevlar. It's used for sports equipment such as kayaks, skis, and helmets. NASA uses Kevlar in ropes for its landing gear.

Firefighter and his gear

Swimwear

Spandex fibers are much different from Kevlar. They are strong but very stretchy. You could stretch a 2-inch piece of spandex to more than 12 inches!

Spandex is stronger than rubber, and it doesn't wear out as fast. But you won't find a label that says "100-percent spandex." Spandex is always blended with other fibers. It's used in athletic clothing, such as swimsuits, bicycle shorts, and dance leotards.

Polypropylene is another textile used for swimwear. It's so light that it floats in water. It dries fast and keeps people warm. It's also used for heavy things, such as carpeting and rope.

be made into clothing. T-shirts and jeans made from these fabrics keep the Sun off people's skin. They are thick and tightly woven or knit. They're often dyed a dark color to protect against sunburns.

SolarMax nylon is used for tents, life jackets, flags, and other things that stay outside in the Sun. They won't rot or fall apart easily.

Future Fibers and Fabrics

Scientists are trying many ways to make new fibers. For example, they hope to create fibers from plants like corn, potatoes, and **cress**. Someday you might have a whole garden of fibers!

Sun-Resistant Clothing

The Sun is very strong. It can burn your skin and rot fabrics. But scientists are making fabrics to beat the Sun. Solarweave nylon, Solarknit cotton jersey, and Solumbra are synthetics that can

TEXTILE Trivia

Critter Hair

- An Angora goat needs to be sheared twice a year. Its mohair grows 1 inch every month. Human hair only grows ½ inch in a month.

Angora rabbit

- Cashmere goats have a soft, fuzzy undercoat. But each goat only grows a small amount of hair that can be used in fabric. It takes hair from 30 goats to make one coat. That's why cashmere fabric is expensive.

- An Angora rabbit grows five or six coats of fur each year. The hairs are 3 to 5 inches long. They are fluffier and warmer than sheep's wool. But the rabbits aren't sheared. The loose fur is plucked off gently. Some rabbits calmly sit in a spinner's lap as the spinner plucks and spins.

Colored sheep have fleece that is jet black, steel gray, silver, fawn, cinnamon, chocolate, or any shade in between. Most colored sheep are born black or brown. Many become lighter as they mature, providing an interesting range of colors.

- Not all wool is white. Sheep can be black or brown. Clean white wool can be dyed any color. Brown or black wool can only be dyed darker colors.

- If the fleece from one sheep isn't washed, a wool sweater will contain ½ cup of oil, a bag of weeds and seeds, and lots of mud!

Growing Fibers

- People listen to the wind to know when to harvest linen. The tiny dry seedpods on the flax stalks rattle when it's windy.

- Linen grows best in cool, rainy weather. Russia and Belgium have ideal climates for this crop.

- A bale of cotton weighs 480 pounds. An acre of land will grow 1½ bales.

Young girl spinning wool at Foster's Mill in England in 1902

Making Fabrics

- Working in the early cotton mills was hard. People worked 15 hours each day. They didn't get paid much either. Children as young as seven worked there.

- Linen was the first fabric people made. It was used to wrap Egyptian mummies 5000 years ago.

- Early knitting machines were a great invention. But the first stockings from these machines didn't have much shape. They were called *leg bags*.

- The first rug was probably made from sheep's wool. People tossed wool on the ground inside their hut to keep the dirt floor dry. But something happened. As people walked over the wool, its fibers matted together. The wool turned into a rug that looked like felt.

Nature's Dyes

- For thousands of years, people used to copy nature's colors. For example, the dark brown stuff in walnut shells dyes fibers brown. So will onion skins. Red berries and beets turn fibers pink or red. In the Midwest, some people use rhubarb leaves to make a yellowish green dye.

Baskets of spices

- Food from the kitchen can be used to dye fibers too. Spices such as turmeric or saffron make yellow dyes. Tea makes brown dyes, and red Kool-Aid makes pink dyes.

- In Mexico and Guatemala, the cochineal bug was used to dye fabrics pink, red, and orange.

- The indigo plant from India was used to dye denim blue.

TEXTILE
Experiments

Exploring Yarn

Untwist a piece of yarn. What do you find? Yarn is made of two or more single strands twisted together. Each strand of yarn is called a *ply*. To make yarn thicker and stronger, more plies are twisted together. Sometimes fancy yarns have shiny or bumpy plies too. If you untwist a single ply, you'll see many tiny fibers.

Let's Try Weaving

Weave a simple mug mat with yarn and a cardboard loom.

Needs

adult help or supervision
cardboard
scissors
ruler
pencil
yarn (2 colors)
straight pin with round head
pencil with eraser

Loom

Procedure

1. Cut a 5-inch square of cardboard.
2. At top and bottom of the square, mark and cut a slit every ¼ inch (19 slits). This will be the warp direction (lengthwise) of your weaving.
3. Wrap yarn around the cardboard, pulling it snugly into slits at top and bottom. Knot ends.
4. Cut a piece of contrasting yarn 4 yards long for the weft. Tie a big knot at one end. Push the pin through the knot and firmly into the pencil eraser. This is your shuttle.

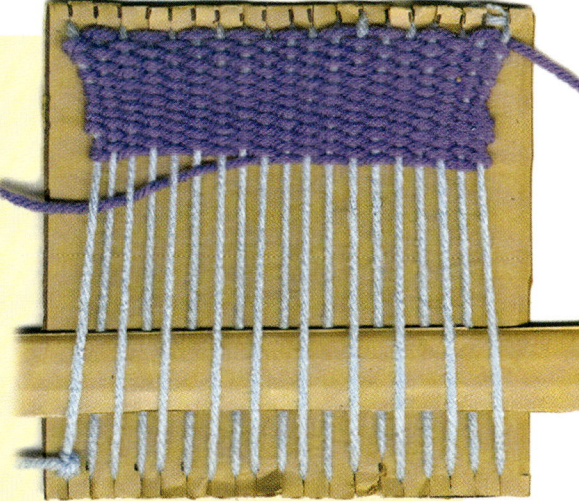

5. Weave from right to left. Push the pencil shuttle under every other warp yarn and over the yarns in between. Push the weft yarn up toward the top of the loom.

6. Bring the shuttle over the last yarn and weave back to the right side. Move the shuttle under all the yarns you went over before. Repeat the patterns.

7. When the yarn is used up, hide the end inside the weaving. Begin a new yarn four or five wefts before the place you ended.

8. When your mat is finished, weave yarn ends back into fabric. Cut yarns in the middle on the back side of the cardboard. Remove the weaving from the cardboard. Tie pairs of yarn edges into knots for fringe.

Tips

- The tighter you push the weft yarns together, the less you'll see the warp yarns.
- Lift the warp yarns with your fingers to help the shuttle pass through.
- Try different weaves.
- Use the pencil tip to push weft yarns together.

Conclusion

We use many different textiles today. We use them when we are awake and even when we sleep. Some are made from natural fibers that people have used for thousands of years. But others are brand-new synthetics that scientists found while experimenting. Textiles are amazing. They keep us comfortable and safe every day.

Internet Connections for *Textiles*

Internet Sites

www.velcro.com/kidzone.html

This site for kids has "The Story of Hook and Loop."

www.cotton.org/pubs/cottoncounts/resources.cfm

Want to learn more about cotton and the consumer? This site offers some interesting facts about one of the most common natural fibers.

www.fibersource.com/f-tutor/history.htm

This site provides a short history of manufactured fibers.

www.fiberworld.com

What do you want to know about fibers and fabric? This site is loaded with information about natural and synthetic fibers.

www.fabriclink.com/University.html

This site for the consumer gives history and facts about fibers, fabrics, and clothing care.

Glossary

artificial (ar tuh FISH uhl) made by humans rather than occurring naturally; made in imitation of something natural

beater (BEE ter) tool that is used to push the weft (see separate glossary entry) threads together

bobbin (BOB uhn) cylinder on which thread or yarn is wound

boll (bohl) rounded seedpod or capsule

card (kard) to comb out and clean wool, cotton, or other fibers (see separate glossary entry) before spinning

cress (kres) plant of the mustard family

crude (krood) not processed

drafting (DRAF ting) pulling along

dye lot (deye laht) fabric, thread, or yarn that is dyed in the same solution so there is no variation in the color

fiber (FEYE ber) fine thread or filament of a natural (see separate glossary entry) or synthetic (see separate glossary entry) material, such as cotton or nylon, that can be spun into yarn

flax (flaks) plant with blue flowers that is widely grown for its seeds, which produce linseed oil, and its stems, which contain the fiber (see separate glossary entry) to make linen

lint (lint) little pieces of thread or fluff

loom (loom) hand- or machine-operated device for weaving thread or yarn into cloth

mill (mil) building used for processing raw materials and manufacturing a product such as fabric or paper

natural (NACH er uhl) present in or produced by nature

pith (pith) central spongy tissue of the stem of some plants

plush (plush) having long, soft fibers (see separate glossary entry) that stick up slightly from the surface of a fabric such as velvet

ret (ret) to soak or moisten plant fibers (see separate glossary entry) such as flax (see separate entry) or hemp so that they become easier to separate

rolag (ROH lag) small roll of fiber that's ready to be spun

shear (shear) to cut hair or fleece from a surface of something using a sharp tool

shuttle (SHUHT uhl) device in weaving that holds the weft (see separate glossary entry) thread and is passed between the warp (see separate glossary entry) threads

spindle (SPIN dul) handheld stick or rod with a notched end through which strands of fibers (see separate glossary entry) are drawn, twisted into thread, and wound around the rod

spinneret (spin er ET) tiny structure that gives out the fluid produced by the glands of a silk-producing worm or spider; device for making filaments of synthetic (see separate glossary entry) fiber (see separate glossary entry), consisting of finely perforated plates through which liquid passes

synthetic (sin THET ic) made artificially by a chemical process

textile (TEKS teyel) cloth or fabric that is woven, knitted, or otherwise manufactured

warp (worp) threads that run lengthwise on a loom (see separate glossary entry) or in a piece of fabric

weft (weft) threads that run horizontally in a piece of woven fabric

Index